We thought that *Nature's Yucky! 2* was a great idea for a children's book, and it seems that people are agreeing with us.

Awards for *Nature's Yucky! 2* *The Desert Southwest*

2008 SKIPPING STONES HONOR AWARD

MOM'S CHOICE GOLD AWARD

USA BOOK NEWS BEST BOOKS AWARD

SOUTHWEST BOOKS OF THE YEAR

What reviewers have to say about *Nature's Yucky! 2*

"Buy this book for the boy who lives in mud puddles or the girl who continually gives some wild creature the comfort of your home. Childhood is an opportunity to explore, and with Nature's Yucky! 2 *the only mess is in your imagination."*

—NEW MEXICO MAGAZINE

"Biology has never been so much fun."
—SOUTHWEST BOOKS OF THE YEAR—BEST READING 2007

"With its "Eeewww! That's Yucky!" refrain, the book is especially fun for reading aloud.

—LEARNING MAGAZINE

NATURE'S Yucky! 2

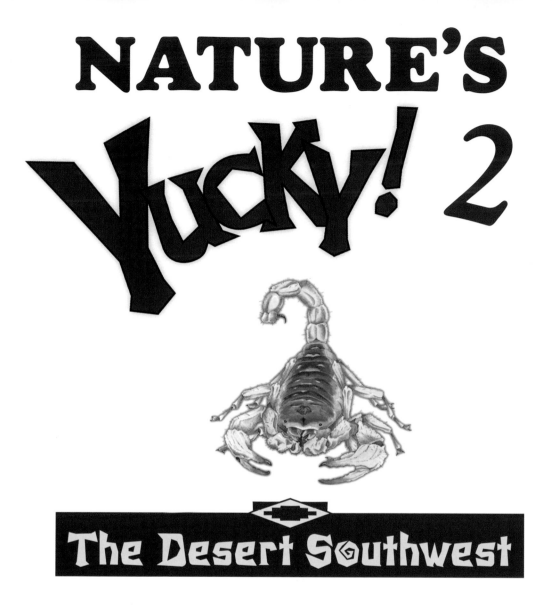

The Desert Southwest

LEE ANN LANDSTROM *and* KAREN I. SHRAGG
Illustrated by Rachel Rogge

Mountain Press Publishing Company
Missoula, Montana
2007

Second Printing, June 2008

AUTHORS' NOTE

As longtime directors of nature centers, we strive to portray nature in a positive light. Some things about nature can be gross, though, no matter what the reasons are for the actions. We hope that teaching children about the benefits of these yucky behaviors will give young people a greater understanding and appreciation of the amazing natural world.

Library of Congress Cataloging-in-Publication Data

Landstrom, Lee Ann, 1954–
    Nature's yucky! 2 : the desert southwest / by Lee Ann Landstrom and Karen I. Shragg ; illustrated by Rachel Rogge.
        p. cm.
    ISBN-13: 978-0-87842-529-7 (pbk. : alk. paper)
    ISBN-10: 0-87842-529-2 (pbk. : alk. paper)
    1. Desert animals—Food—Juvenile literature. 2. Desert animals—Behavior—Juvenile literature. I. Shragg, Karen, 1954– II. Rogge, Rachel, 1971– ill. III. Title. IV. Title: Nature's yucky! two.
QL116.L36 2007
591.754—dc22
2006032479

Printed in Hong Kong by Mantec Production Company

Mountain Press Publishing Company
P.O. Box 2399
Missoula, Montana 59806
(406) 728-1900

To my husband, Jim Heintzman, who shares my passion for language, nature, and travel.

—LEE ANN

To my amazing parents, Babe and Sarah Shragg, on the occasion of their eightieth birthdays, with endless gratitude for their loving support. Half the fun of writing is sharing it with you.

—KAREN

We would like to thank

Lynn Purl, our official editor, for her wordsmithery
and her eagerness for facts and all things yucky

Jim Heintzman, our favorite frontline red-ink editor and fellow science geek

Rachel Rogge for her talent and patience

Rachael Kroog for honoring our books with her clever song

—LEE ANN & KAREN

For my mom and dad, for their unflagging encouragement and faith in me, and for all my friends at the Humboldt State University Natural History Museum, first for welcoming me as a volunteer, and now for continuing to cheer me on.

—RACHEL

Did you know . . .

that **Regal Horned Lizards,**

those prickly, pudgy, prehistoric-looking reptiles,

squirt blood out of their eyes?

Eeewww!!
That's

Yucky!

But hey, it's okay.

Just imagine if it weren't that way!

If it weren't that way, these desert dwellers would have a harder time surviving in their hostile habitat. Horned lizards are a perfect meal for desert hawks, snakes, and even coyotes. Special muscles in the lizard's head stop the blood flow until the increasing pressure bursts tiny blood vessels in and around its eyes. The resulting squirt of blood can reach an enemy as far as three feet away! Using this startling defense may help this camouflaged lizard scare away a predator and live to see another day. If you ever thought it would be fun to stare down one of these reptiles, think again!

Did you know . . .

that **Camel Crickets,**

those heat-loving, humpbacked hoppers,

eat mammal poop?

Eeewww!!
That's

Yucky!

But hey, it's okay.
Just imagine if it weren't that way!

If it weren't that way, camel crickets wouldn't have as much food to eat. Since this insect lives in very dry sand dune areas, where food is hard to find, it eats just about anything: dead or live plant leaves, other insects, and even mammal poop—also known as **scat**. Dry droppings from kangaroo rats, pack rats, jackrabbits, and mule deer are some of the most common foods available in these barren, sandy areas. Scat has leftover food energy in it that the cricket can use. Aren't you glad you don't have to dine on droppings?

Did you know . . .

that **Kangaroo Rats,**

those hardy, hip-hopping hoarders,

have pee like sandy toothpaste?

Eeewww!! That's Yucky!

But hey, it's okay.

Just imagine if it weren't that way!

If it weren't that way, these unusual relatives of mice would not be able to survive in their harsh, dry, desert home. Specialized kidneys allow this nocturnal rodent to get most of the water out of its urine and keep the water in its bloodstream. The leftover waste is a thick goo filled with small, rocky crystals. Its body uses the water it makes digesting dry seeds, which most other animals just pee out as waste. A kangaroo rat is so good at conserving and making moisture that it can survive without drinking water at all when there is none to be found. Makes you thirsty for a tall glass of ice water, doesn't it?

Did you know . . .

that **Sonoran Skinks,**

those shiny, slinky, salamander-shaped lizards,

can break off their tails and leave them

wriggling behind?

Eeewww!!
That's
YUCKY!

But hey, it's okay.

Just imagine if it weren't that way!

If it weren't that way, skinks could not get away as easily from their many enemies. This desert magician can run and disappear after leaving its tail behind. The detached, twitching tail grabs attention while the skink quickly slips away from a bird, snake, or other predator. The skink is able to grow a new, nonbony tail. If you ever tried to catch a skink, you would probably end up with only its tail squirming in your hand—a yucky experience for you and for the skink, because this remarkable reptile would be less likely to escape the next predator without its breakaway tail. So please don't try it. Wouldn't it be wild to have a disposable tail?

11

Did you know . . .

that **Red Satyr Butterflies,**

those delicate, dainty, daytime delights,

sip soupy mud?

Eeewww!!
That's

Yucky!

But hey, it's okay.

Just imagine if it weren't that way!

If it weren't that way, satyrs would not be able to get the minerals they need. This activity is called **puddling**, and it helps the butterfly get salts, minerals, and other nutrients. These are like the nutrients that you get in fortified cereal and chewable vitamins. Male butterflies puddle more often than the females do because the males need more minerals to mate successfully. Next time you are asked to take your vitamins, be glad you don't have to take them in a mud smoothie!

Did you know . . .

that **Western Hognose Snakes,**

those pugnacious, pig-snouted pretenders,

can make themselves smell like garbage?

Eeewww!!
That's **Yucky!**

But hey, it's okay.

Just imagine if it weren't that way!

If it weren't that way, they might get preyed upon more often. Hognose snakes, named for their flat, piglike nose, are famous for their tricks that fool and scare away enemies. A hognose will raise its head, spread its neck like a cobra, and hiss loudly. If that doesn't work, the snake rolls over onto its back, opens its mouth, and plays dead. Then it gives off a bad smell from its cloacal (anal) opening, which may convince a predator to look elsewhere for a meal. Aren't you glad your friends don't do that when you chase them?

Did you know . . .

that **Common Ravens,**

those sleek, soaring, squawking scavengers,

eat the eyes of a dead animal first?

Eeewww!!
That's Yucky!

But hey, it's okay.

Just imagine if it weren't that way!

If it weren't that way, ravens, which do not have a strong, curved beak to rip food like a hawk, would not be able to easily get to a dead animal's meat (**carrion**). Because eyes are soft, the raven can pick at and eat them with its pointed but rather dull bill. The raven gets a quick and easy eyeball meal. Once other animals have opened the carcass, the raven can also eat the leftover body parts. Ravens and other scavengers help recycle dead animals. Even barbeque sauce wouldn't make this meal more appetizing!

17

Did you know . . .

that **Javelinas,**

those pack-loving, piggish peccaries,

smell like stinky feet?

Eeewww!!
That's

Yucky!

But hey, it's okay.
Just imagine if it weren't that way!

If it weren't that way, they couldn't keep the herd together as well. Their bodies make a smelly, skunky, musky oil. A javelina (pronounced HA-vuh-LEE-nuh), also called a collared peccary, stands nose to butt with another member of its pack; they greet one another by rubbing this stinky stuff all over each other. They also rub the oil on nearby plants and rocks to mark their pack's territory. You can often smell javelinas before you can see them! Aren't you glad you can recognize your friends without sniffing them?

Did you know . . .

that **Tarantula Hawk Wasps,**

those tricky tarantula tormentors,

have larvae that eat spiders alive?

Eeewww!! That's  Yucky!

But hey, it's okay.

Just imagine if it weren't that way!

If it weren't that way, the baby wasps would not have a handy, big juicy meal to eat. A mother wasp finds a tarantula burrow and tricks the spider into coming out. There can be a fierce struggle as the wasp tries to sting the spider and paralyze it. The wasp then lays one egg on the tarantula and buries it alive. The spider's own home becomes its underground tomb and the baby wasp's nursery. The larva sucks the juices out of the paralyzed, but still-living, spider. When the baby wasp hatches, it devours the tarantula. Thank goodness people are not on wasp menus!

Did you know . . .

that **White-winged Doves,**

those beige, bobble-headed pigeon cousins, throw up slimy "milk" for their babies?

22

Eeewww!! That's Yucky!

But hey, it's okay.

Just imagine if it weren't that way!

If it weren't that way, the helpless and blind babies wouldn't grow as fast. Both parent doves have a special sac, called a **crop**, in their throats. This crop makes a whitish, milklike liquid full of vitamins, fat, and protein that helps the babies grow quickly. After five days, the parents add seeds to their crops and soften them into a mash with this "pigeon milk." Wouldn't you rather eat oatmeal with cream?

23

Did you know . . .

that **Couch's Spadefoots,**

those secretive, subterranean slumberers,

spend ten months inside a slimy sack?

Eeewww!! *That's* Yucky!

But hey, it's okay.

Just imagine if it weren't that way!

If it weren't that way, these toad cousins couldn't live where they do. During most of the year the desert is dry, with no ponds or puddles, so this unusual amphibian goes into a deep sleep, or **estivation** (like hibernation), underground in a gooey sack. This sack is made of old layers of loose skin and mucus, and smells like roasted peanuts! But don't touch—the slime is toxic to your hands. The spadefoot stays moist and survives inside its slimy sleeping bag until the summer rains start. The spadefoot comes above-ground and eats its skin cocoon for extra energy. After mating, gorging on bugs, and absorbing lots of water, it digs back down into its buried bed-room. Isn't it a good thing you get to use a dry sleeping bag on campouts?

Did you know . . .

that **Circus Beetles,**

those skittery, nighttime scurriers,

squirt smelly liquid from their butts?

Eeewww!!
That's

Yucky!

But hey, it's okay.
Just imagine if it weren't that way!

If it weren't that way, circus beetles would be more easily eaten by other small animals. The circus beetle, also known as a darkling beetle or stink beetle, got its name because of a comical habit. When frightened, it puts its head down on the ground and sticks its butt up in the air, like it's doing a headstand, and spurts a foul fluid that smells like gasoline at an enemy. The spray does NOT wash off but won't hurt people unless it gets in their eyes. Be sure to wear goggles if you want to touch a circus beetle!

27

Did you know . . .

that **Kit Foxes,**

those cunning, clever, curious canines,

lick their babies' butts to make them poop?

Eeewww!!
That's YUCKY!

But hey, it's okay.
Just imagine if it weren't that way!

If it weren't that way, the babies of these small, large-eared relatives of the red fox wouldn't be able to get rid of their waste. The babies' bodies don't completely work just after birth, so the babies, called **kits**, need a bit of help to perform natural functions. Their mother licks them to make them pee and poop. As yucky as this sounds, it is a common practice with mammals such as deer, antelope, and mountain lions. Changing a dirty diaper doesn't seem so bad now, does it?

Did you know . . .

that **Loggerhead Shrikes,**

those fierce, fat-bodied flyers, impale

their freshly killed prey on thorns?

Eeewww!!
That's

Yucky!

But hey, it's okay.
Just imagine if it weren't that way!

If it weren't that way, shrikes, often called butcher-birds, wouldn't be able to eat and store food for a later meal. The shrike is the only songbird that hunts and captures prey like a hawk. A shrike eats lots of insects. But when it catches a mouse, snake, lizard, or bird that is too big to swallow whole, it spikes the prey on a thorn or barbed wire fence to hold it in place while it picks apart and eats it. The spines also hold the food for safekeeping until the shrike returns later to eat. Aren't you glad you can cut your food with a knife and fork?

Did you know . . .

that **Pack Rats,**

those big-footed, bulgy-eyed,

bushy-tailed builders, use pee

to glue their messy

nests together?

Eeewww!! That's  Yucky!

But hey, it's okay.

Just imagine if it weren't that way!

If it weren't that way, pack rats wouldn't have such sturdy homes. A pack rat, or wood rat, gathers lots of stuff to make its nest. It takes fanciful, shiny items from people (such as bottle caps, bullet shells, coins, keys, jewelry, small toys, and eyeglasses), as well as twigs, leaves, pine needles, nuts, cactus chunks, reptile scales, dried poop, and even bone bits, to form a pile up to two feet high and four feet wide, with tunnels inside. The rodent's sticky urine acts like glue to hold everything together. The thick walls protect the rat and keep it cool. Aren't you glad that people use wood, stone, or brick to build their houses?

Did you know . . .

that **Dung Beetles,**

those daring, daunting decomposers,

really do eat animal poop?

Eeewww!!
That's
Yucky!

But hey, it's okay.

Just imagine if it weren't that way!

If it weren't that way, nature would lose one of its best cleanup crews, the soil would be less fertile, and baby dung beetles, or **larvae**, wouldn't have a ready meal when they hatch inside their dung-ball nursery. The adult male beetle rolls a ball of **dung** (a fancy name for poop) as an attractive present for females. When he finds a willing mate, they bury the dung ball and she lays her eggs inside it. When the eggs hatch, the larvae munch on the nutrient-rich dung. Decomposers like dung beetles show that everything in nature is recycled—even disgusting poop. When you think about what dung beetles feed their kids, maybe you won't complain about what you're having for dinner!

More Fun Facts About
Nature's Yucky! Desert Animals

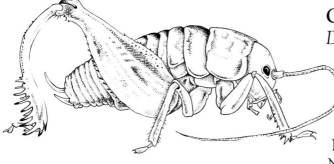

Regal Horned Lizard
Phrynosoma solare

Size: 5½ inches (14 centimeters)

Food: prefers ants, but also eats grasshoppers, beetles, and spiders

Range: Sonoran Desert in southern Arizona and New Mexico, south into Mexico

The horned lizard is often mistakenly called a "horned toad" or "horny toad." Toads are amphibians, while horned lizards are reptiles with a very different skin type and life cycle. Only 4 of the 14 species of horned lizards can squirt blood. Foxes, coyotes, hawks, roadrunners, snakes, and other lizards eat horned lizards, despite their horns and camouflaged, scaly skin. Humans threaten horned lizards' survival by moving to desert regions, putting up buildings, and poisoning ants in their yards, thus eliminating the reptiles' major food source.

Giant Sand Treader Camel Cricket
Daihinibaenetes giganteus

Size: 1½ inches long (4 centimeters)

Food: small bits of dead plants, insect bodies, poop

Range: scattered sites in Colorado, New Mexico, and Utah

Giant sand treader camel crickets were once thought to live only in Great Sand Dunes National Park in southern Colorado, but two other closely related camel cricket species have been found elsewhere. The camel cricket makes its home in sand dunes and dry streambeds. Named for its humped back, this large, wingless cricket feeds at dusk. It has special hairs and spurs called sand baskets on its feet, making it an excellent digger. It absorbs moisture from the sandy soil in its burrow, which can be up to 18 inches long.

Ord's Kangaroo Rat
Dipodomys ordii

Size: body, 8 to 11 inches (20 to 28 centimeters); tail, 4 to 6½ inches (10 to 17 centimeters); 6 to 7 ounces (170 to 200 grams)

Food: seeds of various desert shrubs, grasses, other flowers; sometimes green vegetation

Range: widespread in western North America from southern Canada to Mexico

The kangaroo rat hops like a kangaroo, using its tail for balance. The soles of its feet are hairy, which helps give it traction in loose sand. It survives the hot desert days and avoids losing moisture by staying in its comfortable underground burrow and coming out at night to scavenge for seeds, which it **caches** (stores) for later use. It does not sweat or pant. It is mostly solitary and has a small territory of about half an acre. Mothers usually have 2 babies per litter and 2 or 3 litters per year. Kit foxes and snakes are the major predators. Even though it is efficient in its adaptation to desert climes, the kangaroo rat is listed as a species of special concern because its population is in decline from habitat loss.

Sonoran or Great Plains Skink
Eumeces obsoletus

Size: 6½ to 13¾ inches (16.5 to 35 centimeters)

Food: grasshoppers, crickets, spiders, moths, butterflies, and insect larvae or nymphs

Range: Arizona, New Mexico, and western Texas north to the Great Plains

This lizard likes grasslands with fine soil to burrow in. In deserts it lives along waterways or semipermanent water sources. It is secretive, hiding under rock slabs, so it is rarely seen. If a predator grabs the skink, its tail can break off in a weak area of the tailbone (vertebrae). The tail stump doesn't bleed much because sphincter muscles close off nearby arteries. The tail does grow back, but it grows back with cartilage in place of the bones. Sometimes an extra tail grows, so you could find a two-tailed or even a three-tailed skink!

Red Satyr Butterfly
Megisto rubricata

Size: 1⅜ to 1⅞ inches (3.5 to 4.7 centimeters)

Food: flower nectar

Range: central Arizona and New Mexico, to eastern Texas and central Kansas, and south to Guatemala

Red satyr butterflies prefer open brushland or oak-pine forests. The male patrols shady areas, looking for females. The female lays eggs on dead leaves or near grass blades. The caterpillar eats grasses, especially Bermuda or St. Augustine grass; the mature caterpillar hibernates before awakening in the spring to eat, pupate, emerge as an adult, and start the cycle all over again. The red satyr has 2 **broods** (sets of babies) from April to December. Like many insects, and butterflies in particular, it is important in the food chain because it pollinates flowers and serves as food for birds. As the world becomes more paved over, butterflies of all types are on the decline.

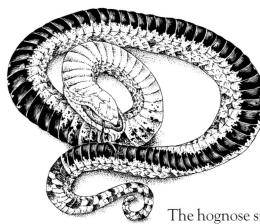

Western Hognose Snake, Mexican subspecies
Heterodon nasicus kennerlyi

Size: 15 to 25 inches (38 to 63.5 centimeters)

Food: primarily toads, but also frogs, lizards, small birds, small mammals, insects, reptile eggs

Range: this subspecies is found in the southern portions of Arizona, New Mexico, and Texas

The hognose snake has a slightly toxic saliva that flows down a pair of chewing fangs in the very back of its mouth. These fangs can also puncture a puffed-up toad, making it easier to swallow. This snake hunts in the early morning and late evening; it grabs and squeezes its prey to suffocate it, then swallows it whole. The western hognose is a good burrower, which helps it escape predators and find toads. This snake prefers sandy, dry prairie and grassy areas. The female (which is larger than the male) typically lays 15 to 35 eggs, 1½ inches (4 centimeters) long, which hatch after 50 to 64 days.

Common Raven
Corvus corax

Size: 22 to 37 inches (56 to 94 centimeters); 2 to 3½ pounds (1 to 1.5 kilograms)

Food: grain, berries, seeds, eggs, frogs, mice, carrion (dead animals)

Range: widespread in North America, Eurasia, and Africa

The raven is the largest member of the crow family. The male and female look alike, but the female is a bit smaller. The female incubates the eggs, and both parents feed the 4 to 6 young and carry water to them in their throats. Even by human standards, ravens seem smart. They easily adapt to new situations and exploit new food sources. Many indigenous people revere ravens for their cleverness, and there are many traditional stories about ravens. Because they are such good scavengers, they are important recyclers in nature. Unfortunately, ravens often die when they eat the carcasses of animals that have been poisoned by humans.

Javelina (Collared Peccary)
Tayassu tajacu

Size: 21 to 24 inches tall (53 to 61 centimeters); 30 to 40 inches long (76 to 102 centimeters); 40 to 65 pounds (18 to 29 kilograms)

Food: roots, cactus (particularly prickly pear), acorns, seeds, sometimes insects or dead meat

Range: southern Arizona, New Mexico, and Texas, south to South America

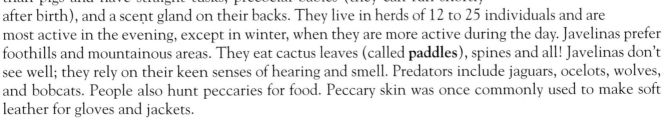

Although they look like pigs, javelinas are peccaries. Pigs are European animals, while peccaries are from the New World. Peccaries are smaller than pigs and have straight tusks, precocial babies (they can run shortly after birth), and a scent gland on their backs. They live in herds of 12 to 25 individuals and are most active in the evening, except in winter, when they are more active during the day. Javelinas prefer foothills and mountainous areas. They eat cactus leaves (called **paddles**), spines and all! Javelinas don't see well; they rely on their keen senses of hearing and smell. Predators include jaguars, ocelots, wolves, and bobcats. People also hunt peccaries for food. Peccary skin was once commonly used to make soft leather for gloves and jackets.

Tarantula Hawk Wasp
Pepsis species

Size: up to 2 inches long (5 centimeters)

Food: adults eat flower nectar and pollen, favoring yucca blossoms; larvae eat tarantula spiders

Range: Chihuahuan Desert and surrounding regions

Often simply called tarantula hawks, 9 species of tarantula hawk wasps live in the desert Southwest. The reddish brown or orange wings are a warning to predators, since this wasp has one of the most painful stings of any insect. People who have been stung report the pain is unbearable, but it does not last long, and the sting is not especially lethal. The female's stinger is up to ⅓ of an inch (0.8 centimeters) long. Roadrunners are one of the few predators. Males perch on top of tall plants, watching for females and defending their territory. The tarantula hawk wasp is the official state insect of New Mexico, nominated by schoolchildren.

White-winged Dove
Zenaida asiatica

Size: 11 to 12½ inches long (28 to 32 centimeters); 4½ to 7 ounces (127.5 to 198.5 grams)

Food: saguaro cactus nectar and fruit, other seeds and fruits, acorns

Range: Texas, Arizona, Nevada, New Mexico, and southern California, south to Panama

This dove is well adapted to the desert. It can survive 5 days without drinking and can tolerate losing 20% of its body's water weight. It can also tolerate water that is 25% salty. It secretes its urine as an acid, not urea, which conserves a lot of water. Proportionally, it can drink 10 times more water at once than a person can. Often this bird can get all the water it needs from saguaro cactus fruit. The female usually lays 2 eggs in a twiggy nest; the eggs are incubated by both parents. The chicks leave the nest after 13 to 15 days. These doves can "store" excess body heat (4 to 6 degrees) during the day and release it during the cooler night. In winter, they migrate to Central America.

Couch's Spadefoot
Scaphiopus couchii

Size: 2 to 3 inches (5 to 7.5 centimeters)

Food: tadpoles eat microscopic organisms, algae, filtered organic bits; adults eat fairy shrimp, aquatic weeds, dead tadpoles, plant matter, termites that wash into the pond

Range: Great Basin, southwest Oklahoma to southeast California, south into Mexico

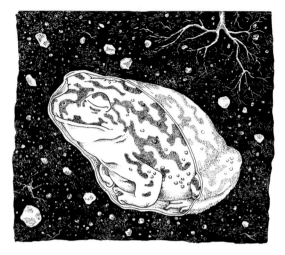

Like all amphibians, the spadefoot needs water to breed. After awakening from estivation, the male calls loudly, sounding like a bleating sheep, to attract a female. They use natural pools, as well as storm runoff basins, roadside ditches, and puddles in vacant lots, to breed and lay eggs. Tadpoles hatch and grow into adults in just 9 to 14 days, before the pools dry up. The spade on each hind foot is a dark, hard projection that helps them dig down over 2 feet into the ground. The noise of ATVs (all-terrain vehicles) and motorcycles sounds like thunder, which causes the toads to wake up early and use up their stored energy, only to find a dry desert with no rain.

Circus Beetle
Eleodes obscurus

Size: up to 1½ inches long (3.3 centimeters)

Food: tiny pieces of dead leaves, dead insect parts, fungus bits

Range: throughout the southwestern deserts

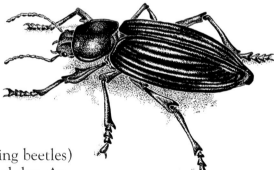

Circus beetles (also known as pinacate, clown, or darkling beetles) are easily seen, since they wander around both night and day. Another nickname, "stinkbugs," comes from their defensive stinky spray, which protects against predators such as scorpions, tarantulas, and ants. Desert mice avoid the spray, however, by grabbing a beetle, sticking its abdomen in the soil, and eating the top half of the beetle. Skunks and burrowing owls also eat them. Circus beetles can live in the driest habitats. Tiny belly hairs collect moisture from the cool night air, which runs down the belly to the beetle's mouth. The larvae live in the soil or dead leaves and eat dead plants.

41

Kit Fox
Vulpes macrotis

Size: body, 15 to 20 inches (38 to 51 centimeters); tail, 9 to 12 inches (23 to 30.5 centimeters); 3 to 6 pounds (1.4 to 2.7 kilograms)

Food: small rodents

Range: southwestern U.S. and northern Mexico

Kit foxes have extraordinarily large ears in order to hear better and to get rid of extra body heat. This fox prefers open sandy areas or low bushy areas; they are the only North American fox that can live in the extreme desert environment. They stay in their cool burrows during the day and hunt at night. Four to seven pups are born in the early spring. The kit fox is the second smallest fox in the world and smaller than an average house cat. Unfortunately they are becoming rarer as increasing numbers of people move into their limited habitat and convert the landscape into houses, stores, and farm fields. Kit foxes are helpful to humans because they eat a lot of rodents, but they are often the unintended victims of poison put out for coyotes.

Loggerhead Shrike
Lanius ludovicianus

Size: 9 inches (23 centimeters); 1½ to 2 ounces (42.5 to 56.5 grams)

Food: insects, especially grasshoppers and bumblebees, small mammals, small birds, lizards

Range: year-round in the southern half of the U.S.; summer breeder in the northern plains and mountain states

The loggerhead shrike is a robin-size songbird that hunts like a bird of prey. The shrike has a hooked bill and tremendous eyesight. It mainly eats insects in the summer and small songbirds and mice in the winter. Like hawks and owls, it even coughs up "pellets" of bone and fur, depending on what it has eaten! Shrikes like open fields with scattered trees or fences where they can perch and watch for prey. Loggerhead shrikes are slightly smaller than northern shrikes, but they look so much alike that identification is difficult, even for experienced birders. Loggerhead shrike populations have been suffering serious decline throughout North America as land is developed for housing and commerce.

White-throated Wood Rat (Pack Rat)
Neotoma albigula

Size: 12 to 16 inches long (30.5 to 40.5 centimeters), including the tail; 5 to 10 ounces (135 to 283 grams)

Food: mostly cactus leaves; also seeds, mesquite beans, yucca leaves, grasses

Range: Sonoran and Chihuahuan deserts in Arizona, New Mexico, and adjacent states

Pack rat nests, called **middens**, are natural time capsules; some nests are over 40,000 years old! By looking at the almost-fossilized bone bits, seeds, plant matter, insect parts, pine needles, and pollen inside, scientists can tell how the habitat and climate have changed over time. The nests are usually placed at the base of cholla cactus and are full of collected "goodies," including cactus pads. These spiny fortresses help protect the wood rat from predators. The wood rat is active at night. Often the only water it gets comes from the cactus it eats.

Dung Beetle
Canthon species

Size: 1 to 2 inches (2.5 to 5 centimeters)

Food: scat, decaying vegetation, and fungi

Range: Arizona, New Mexico, and east of the Rockies

These beetles eat more than their own weight in poop each day. This serves a significant role in creating good soil. Dung beetles recycle the droppings of other animals and fertilize the soil by mixing in the rich organic matter. They help aerate the soil when they dig into the hard ground to bury their dung balls. When they roll their balls around, the beetles also disperse seeds to new locations. Desert dung beetles compete with each other for quick-drying and hard-to-find mammal scat. The scarab beetle, a type of dung beetle, was considered holy in ancient Egypt because it seemed to come alive again (be resurrected) after it went underground to bury its dung balls.

Raven-gut Upside Down Cake

This a great-tasting dessert that demonstrates how the guts of ravens are full of seeds from the berries they eat, which get spread every time they poop.

1 angel food cake (either homemade or store-bought), cut into one-inch slices

3 15-ounce cans of soft fruit, drained (blueberries, blackberries, and pitted cherries work best. You can also use peaches or nectarines.)

$1/2$ cup nuts, such as chopped walnuts, sliced almonds, or sunflower seeds (optional)

2 tablespoons of sugar

Grease a medium-sized ceramic bowl with butter or margarine. Line the bottom and sides with half the cake slices.

Fill the middle of the bowl with 3 cans of drained berries and the nuts. Add sugar to taste.

Cover the fruit-filled bowl with the remaining slices of cake.

Put a heavy plate on it to compress the contents. Refrigerate overnight.

Loosen the edges with a knife. Invert the bowl onto a platter and slice the cake.

It's delicious! Although messy.

We recommend that you buy organic ingredients whenever possible. This is another way to help wildlife, the soil, and you. Organic food products are helpful since they reduce pesticides and herbicides on farmland and in water.

How do I begin to write a song about this wonderfully disgusting book [the first *Nature's Yucky!*]? Well, obviously I needed to go to the same level that Karen and Lee Ann did when they wrote it. So . . . here it is!

NATURE'S YUCKY! BLUES

By Rachael Kroog

If you're looking for a book that's really, really gross,
It's rotten, more rotten than most—*Nature's Yucky!*
Meet Lee Ann and Karen and you'll get lucky,
'cause they just wrote a book, it's called *Nature's Yucky!*
Nature's Yucky!
It's disgusting, but hey, that's okay.
Just imagine if it weren't that way!
Nature's Yucky!

It can be long, it can be short, it can be round or fat.
You can call it animal droppings, poop, or scat.
Did somebody say scat? How 'bout that, yeah?
Zip-a-dee-doo-dah, zip-a-dee-doo-doo.
It's made of berries and fur, it's made of twigs,
it's made of hair.
Animals, they don't wear no underwear!

Nature's Yucky!
It's disgusting but hey, that's okay.
Just imagine if it weren't that way!
Nature's Yucky!

Learn some nasty stuff about our animal friends.
Page after page, the "yucky" never ends.
From moose to salmon, to bears to owls,
These fascinating facts are filled with foul (fowl).
It goes in one end and out the other.
If you don't know what I mean,
you'd better ask your mother!
Nature's Yucky!
It's disgusting, but hey, that's okay.
Just imagine if it weren't that way!
Nature's Yucky!

To hear Rachael sing her song, go to www.rachaelkroog.com and click on KidPower, then Audio.

RESOURCES

ADULT BOOKS

Arritt, Susan. *The Living Earth Book of Deserts*. Pleasantville, NY: Reader's Digest Association, 1993.

Broyles, Bill. *Our Sonoran Desert*. Tucson: Rio Nuevo Publishers, 2003.

Larson, Peggy Pickering, and Lois Larson. *The Deserts of the Southwest: A Sierra Club Naturalists' Guide*. San Francisco: Sierra Club Books, 1990.

Lazaroff, David. *Arizona-Sonora Desert Museum Book of Answers*. Tucson: Arizona-Sonora Desert Museum Press, 1998.

Oldfield, Sara. *Deserts: The Living Drylands*. Cambridge, MA: The MIT Press, 2004.

Tweit, Susan J. *Seasons in the Desert: A Naturalist's Notebook*. San Francisco: Chronicle Books, 1998.

Zwinger, Ann Haymond. *The Mysterious Lands: A Naturalist Explores the Four Great Deserts of the Southwest*. Tucson: University of Arizona Press, 1996.

JUVENILE NONFICTION

Davies, Nicola. *Extreme Animals: The Toughest Creatures on Earth*. Cambridge, MA: Candlewick Press, 2006.

Jablonsky, Alice. *101 Questions about Desert Life*. Tucson: Western National Parks Association, 1994.

Johnson, Rebecca L. *A Walk in the Desert*. Minneapolis: Carolrhoda Books, 2001.

MacQuitty, Miranda. *Desert* (Eyewitness Books). New York: DK Publishing, 2000.

Masoff, Joy. *Oh Yuck! The Encyclopedia of Everything Nasty*. New York: Workman Publishing, 2000.

Rozario, Paul. *Spreading Deserts*. Chicago: Raintree, 2004.

Savage, Stephen. *Animals of the Desert*. Austin, TX: Raintree Steck-Vaughn, 1997.

Steele, Christy. *Desert Animals*. Austin, TX: Raintree Steck-Vaughn, 2002.

Wallace, Marianne D. *America's Deserts: Guide to Plants and Animals*. Golden, CO: Fulcrum Publishing, 1996.

PICTURE BOOKS

Baylor, Byrd. *Desert Voices*. New York: Aladdin, 1993.

Bessesen, Brooke. *Look Who Lives in the Desert! Bouncing and Pouncing, Hiding and Gliding, Sleeping and Creeping*. Phoenix: Arizona Highways Books, 2004.

Cronin, Doreen. *The Diary of a Worm*. New York: Scholastic, 2003.

Fredericks, Anthony D. *Around One Cactus: Owls, Bats, and Leaping Rats*. Nevada City, CA: Dawn Publications, 2003.

———. *Under One Rock: Bugs, Slugs, and other Ughs*. Nevada City, CA: Dawn Publications, 2001.

Lowell, Susan. *The Three Little Javelinas*. Flagstaff, AZ: Rising Moon Press, 2003.

Pallota, Jerry, and Ralph Masiello. *The Yucky Reptile Alphabet Book*. Watertown, MA: Charlebridge Publishing, 1989.

WEB SITES

Sites about desert animals:

www.desertusa.com/animal.html
Very thorough site.

www.desertmuseum.org
Arizona-Sonora Desert Museum site. Go to Kids & Education/Creature Features.

www.gf.state.az.us
Arizona Game and Fish Department.

www.nps.gov/grsa
Great Sand Dunes National Monument site.

www.friendsofsaguaro.org
Saguaro National Park. Lots of desert information. Click on Kid's Page for fun facts and activities.

www.nps.gov/moja
Mojave National Preserve. Go to Nature & Science to explore the animals, plants, and terrain.

http://wc.pima.edu/~bfiero/tucsonecology/index.htm
Pima Community College site on the desert ecology of Tucson.

www.mnh.si.edu/mna/search_name.cfm
Smithsonian National Museum of Natural History's searchable Web site of North American mammals.

www.enchantedlearning.com/biomes/desert/desert.shtml
Get animal information and pictures to print out and color.

http://bugguide.net
Searchable database of hundreds of insect species. Check out the tarantula hawk wasp photos!

http://entweb.clemson.edu/museum
Clemson University's searchable "cabinet" of insects and photos.

www.scarabnet.org
A beetle database, searchable by species, genera, or distribution; some photos.

Organizations that help wildlife:

www.audubon.org
National Audubon Society

www.nwf.org
National Wildlife Federation

www.nature.org
The Nature Conservancy

www.wcs.org
The Wildlife Conservation Society

www.iwla.org
The Izaak Walton League

About the Authors

Lee Ann Landstrom and **Karen I. Shragg** are longtime educators of children and adults. Landstrom has a master's degree in biology and directs the Eastman Nature Center for the Three Rivers Park District in Osseo, Minnesota. Shragg has a master's degree in outdoor education and recreation, and a doctorate in education. She heads the Wood Lake Nature Center in Richfield, Minnesota. Their first book together, *Nature's Yucky! Gross Stuff That Helps Nature Work*, received the Izaak Walton League of America Book of the Year Award.

About the Illustrator

Rachel Rogge studied studio art and biology at Humboldt State University, where she also worked at the Natural History Museum. She holds a graduate certificate in science illustration from the Science Communication program at the University of California, Santa Cruz, and is a freelance illustrator. This is her first children's book.

Additional Books for Young Readers

_____Awesome Ospreys: Fishing Birds of the World	$12.00	For ages 8 and up
_____Bold Women in Michigan History	$12.00	For ages 12 and up
_____The Charcoal Forest: How Fires Help Animals and Plants	$12.00	For ages 8 and up
_____Loons: Diving Birds of the North	$12.00	For ages 8 and up
_____Medicinal Plants of North America: A Flora Delaterre™ Coloring Book	$12.50	For ages 7 and up
_____Nature's Yucky! Gross Stuff That Helps Nature Work	$10.00	For ages 5 and up
_____Nature's Yucky! 2: The Desert Southwest	$12.00	For ages 5 and up
_____Owls: Whoo are they?	$12.00	For ages 8 and up
_____Sacagawea's Son: The Life of Jean Baptiste Charbonneau	$10.00	For ages 10 and up
_____Shanleya's Quest: A Botany Adventure for Kids Ages 9 to 99	$12.50	For ages 9 and up
_____Shanleya's Quest: Patterns in Plants Card Games	$12.50	For ages 9 and up
_____Snowy Owls: Whoo Are They?	$12.00	For ages 8 and up
_____Spotted Bear: A Rocky Mountain Folktale	$15.00	For ages 5 and up
_____Stories of Young Pioneers: In Their Own Words	$14.00	For ages 10 and up

Please include $4.00 shipping and handling for 1–4 books and $6.00 for 5 or more books.

Send the books marked above. I have enclosed $_____

Name_____Phone_____

Address_____

City/State/Zip_____

☐ Payment enclosed (check or money order in U.S. funds)

Bill my: ☐ VISA   ☐ MasterCard   ☐ American Express   ☐ Discover

Card No._____Exp. Date_____

3 Digit Security Code_____Signature_____

MOUNTAIN PRESS PUBLISHING COMPANY
Post Office Box 2399 / Missoula, Montana 59806
PHONE 406·728·1900 / FAX 406·728·1635 / TOLL FREE 1·800·234·5308
E-MAIL info@mtnpress.com / WEBSITE www.mountain-press.com